ROBERT DRUMMOND, PRINTER, NEW YORK.

PREFACE.

This book is intended, by a representative course of progressive exercises found in the two hundred hour Course in Drawing in the first part of the book and by a systematically arranged compilation of precept found in the Manual of Drawing in the second part of the book, to convey the essentials of modern conventional drafting as practiced by the general profession of mechanical engineering.

The Course in Drawing is divided for convenience into five grades: The drawings of the first grade are to be practically copies of the four assigned plates and are to be made from detailed instructions in procedure.—The object of the work in this grade is to give the student the science of the elementary principles of drafting and instruction in the use of the materials and instruments and in the technic involved. The drawings of the second grade are to be detailed working drawings made from sketches of machine parts.—One of the most important acquirements for a draftsman is to be able to sketch machine details neatly, accurately and with facility; and, for a student, hardly anything is more interesting and instructive than working from something that has definite form and utility and, incidentally, a large amount of the form of good design may be acquired by absorption. The drawings of the third grade are to be assembly drawings made from sketches of machine parts. The drawings of the fourth grade are to be detailed working drawings and detailed assembly drawings from plates which are copies of commercial drawings.—

It is excellent practice for a student to an machines and their parts in order to acquire mind the shape and proportions of the piece the fifth grade are foundation drawings wit plans, and Patent Office drawings.—These dr by themselves, and conventions employed in understood by all well equipped draftsmen, an involved in Patent Office drawing is invalua be layman or draftsman.

The drawing course has been extended t complete detail drawings of a 15" Pillar Sha making assembly drawings. These plates ar mercial machine

The Manual of Drawing describes the materials and instruments and all the essent execution of a complete commercial drawing. is to put into definite form a single and star which shall be the average of the systems us facturing concerns. For this purpose, data one hundred and thirty of the largest concer various lines of business.

Illustrations by line-cuts and half-tones ar

the Manual to make clear the important points discussed in the text. In addition, twenty-one 9×12 zincograph plates are bound into the back part of the book to accompany the Course in Drawing.

An 8×11 pocket in the inside of the back cover of the book is provided for holding loose sketches and any detached material that may be used in connection with the book.

This book is specifically written for the use of the students in drawing in the present Sophomore class of Sibley College, for students in manual training schools, trade schools, and technic draftsman who is serving his apprenticesh

The authors desire to thank all the m; so kindly lent assistance by contributing

The first nineteen zincograph plates in made from copies of drawings executed b;

ITHACA, N. Y., September, 1904.

COURSE IN DRAWING.

COURSE IN DRAWING.

INTRODUCTION.

SUPPLIES REQUIRED FOR THE COURSE IN DRAWING.

6 sheets 18×24 Keuffel & Esser's Duplex (or equivalent) Drawing Paper.

6 sheets 18×24 Crane's (or equivalent) Bond Paper.

1 sheet 18×24 Paragon (or equivalent) Drawing Paper.

2 sheets 18×24 Imperial Tracing Cloth.

1 sheet 10×15 two-ply Bristol Board.

8×10 book of Cross Section Paper.

$\frac{3}{4}''×20''×26''$ soft pine Drawing Board.

26″ pear-wood T Square, head fixed with five screws.

10″ 30°×60° transparent Triangle.

7″ 45° transparent Triangle.

12″ triangular Boxwood Scale (Sibley scale), graduated with scales as follows: $\frac{1}{8}''$, $\frac{1}{16}''$, $\frac{1}{4}''$, $\frac{3}{8}''$, $\frac{3}{4}''$, $\frac{1}{2}''$, $1''$, $1\frac{1}{2}''$, $3''$, $6''$, and $12''$ to the foot.

5½″ Compass with fixed leg for needle point, detachable pencil- and pen-legs and lengthening bar.

5″ Dividers with hair-spring adjustment.

Bow Pen, Pencil and Spacer for scribing not over $1\frac{1}{2}''$.

5″ Unhinged Ruling Pen.

A pocket case for instruments.

Scale Guard.

Beam Compass Clamps and Beam.

Transparent Logarithmic Spiral Curve—the equivalent of Keuffel & Esser's, Catalog No. 1861.

1 set of Hardtmuth's HHHHHH leads for Comp
1 HHH and 1 HHHHHH Hardtmuth's Koh-i-n
Good Pencil Pointer. Sandpaper pad or 10″ flo
Ink Eraser—Faber's Typewriter.

Pencil Eraser—Tower's Multiplex.

Package of 1 oz. Copper Tacks or a card
Tacks.

Arkansas Knife Oil Stone.

Pen Staff and ball-pointed Writing Pen.

Bottle of black waterproof Drawing Ink.

6″ Machinist Scale.

6″ Caliper.

The following Special Instruments of better and useful attachments, can be substituted for ar have identical use:

26″ Mahogany T Square, head fixed with five ebony lined head and blade, and with workin (Get no transparent lined edges.)

8″ transparent combination Triangle, D. J. Kel Conn.

3 12″ sheet steel, nickel plated, Brown & Sharpe graduations as found above on the Boxwood

5″ first quality, and first class designed hinged pricker point in handle.

PROOF AND PREPARATION OF INSTRUMENTS.

Wipe all instruments clean with a lintless rag—chamois skin is best.

Prove the straightness of the working edge (the left hand edge is the working edge for a right hand person.—**Fig. 1.***) of drawing board by applying a standard straight-edge or the edge of a T square blade if it has been proved straight.

Prove the top surface of the drawing board for convexity or planeness by applying a standard straight-edge, or the edge of a T square blade to its surface.

Prove the squareness of the working edge of the drawing board to its lower edge by the following method: First, place T square head against working edge of board and draw two parallel lines, one about one inch from the upper edge of board, and the other about one inch from the lower edge of board. Second, with the T square head placed against the working edge of board, mark off on the T square blade a point whose distance from the edge opposite the working edge of board will be about the length of the T square head. Third, with the T square head placed against the working edge of board, sight and mark off from the point on the T square blade a corresponding point on each of the parallel lines. Fourth, place the T square head against the lower edge of board and draw a line through the point marked off on one of the parallel lines and note if it passes through the corresponding point on the other line. Absolute squareness is unnecessary.

Prove the straightness of the upper edge (the only edge used) of the T square blade by the following method: First, draw with it a line. Second, turn it end for end and place the edge of the blade to exactly coincide with each end of the line which was drawn. Third, draw another line. Fourth, note the coincidence of the lines drawn.

Prove the straightness of the T square head by applying a standard straight-edge or the long side of a triangle.

Prove the 90° angles of the triangles by the following method: First,

place the short side of the triangle again and draw a line with the 90° side of the t over and with same side of triangle dr first and note the coincidence of the lin

Prove the 45° angles of the triangle place one of the short sides of the triangl blade and draw on paper with a sharp point with the 45° side of the triangl side of the triangle against the edge another line through the same point w Third, draw a horizontal line across drawn. Fourth, measure with divider isosceles triangle.—This should follow tl

Prove the 30° and 60° angles of the First, place the short side of the triangle blade and draw a line through a point Second, turn the triangle over and with s line through the same point. Third, extremities of the lines just drawn. F inclined lines and note the equilateral tr

Prove the scale by the following metl line drawn on paper its equal subdivis scale end for end and note the coincider

Set into the pencil leg of the compass it to project not more than $\frac{1}{4}''$. Sharper to an edge that is slightly elliptical. D and forth for a short distance across the it about its axis. Set needle point into the pencil point so that, when both a the needle point and point of the pencil

If needle points of compasses, divid from ordinary wear, they should be g are sharp enough to offer a slight resistance to being withdrawn from the paper into which they are stuck under very slight pressure.

Set needle points and "lead" into bow instruments in the same manner as described for the compass.

Repair the ends of the nibs of the ruling pen by the following method: First, clean the pen thoroughly. Second, hold the pen to the light or over a piece of white paper and close the nibs until they are about to touch. Third, hold the pen with its axis perpendicular to the plane of the knife oil stone and move the stone over the points of the nibs until an arc of not over $\frac{1}{64}''$ radius is rounded on the points.—**Fig. 6.** Fourth, open the blades as wide as possible. Fifth, hold the axis of the pen at right angles with a line along the length of the stone and, with one blade at an angle of about 30° with its plane surface (**Fig. 7**), move the point of the nib about one half inch back and forth along the stone while the pen is being slightly oscillated about its axis to maintain the roundness on the back and grind the point of the nib to an edge as sharp as a pocket pen-knife.—This operation should be performed on each nib separately and with extreme care to avoid grinding off the end of a nib, thus making one nib shorter than the other. Sixth, rub lightly, with the thin edge of the stone, the inside surface of each nib to remove any possible burr. Seventh, test the pen thus: clean nibs thoroughly; close the nibs until they are about to touch; fill with ink to a depth of not over $\frac{3}{8}''$; hold the pen with its axis slightly inclined to the right and in a plane perpendicular to the plane of the paper and draw it lightly from left to right over drawing paper until the ink is exhausted (**Fig. 8**) and note that the pen never fails to make a line while it is moving across the paper and that the line is clean cut.

If the ink in the drawing pens refuses to flow, draw the pen once or twice across the little finger and, if this does not avail, clean the pen and refill.

Clean all pens the instant they cease to be

Sharpen both ends o. both pencils, one end apex is slightly blunted, and the other to a wedge $\frac{3}{8}''$ from the point and whose edge is slightly "lead" should project from the wood not less th $\frac{1}{4}''$.—**Fig. 5.**

PRINTING.

Plate I.—Standard Printing, Lines, Form c of Title.

Tack to the drawing board a sheet of 18×2 with smooth side up.

Print in title shown on Plate I.

All letters and figures used on commercial simple type, plain, of uniform height and com width and of the same width as the visible line the drawing. The inclined straight line printi Plate I is recommended for use on the drawings, a plainest, and usually more easily done than any

Draw parallel guide lines with the pencil b figures.

The writing pen, when used for printing, sho from its point than $\frac{1}{4}''$. The farther away the more unsteady will be the pen and consequentl be the printing.

Crowd the letters of a word close together an

Make the initial letters of all words excej junctions a little taller than the others.

Do not make figures or letters less than $\frac{1}{16}''$

FIRST GRADE.

Make finished inked drawings of Plates II, III, IV, and V. These drawings should be exact copies of the arrangement of every single detail within the margin lines of the plates.

Plate II.—Driving Stud, Worm, and Carriage Feed Screw and Plunger.

Tack to board sheet of 18×24 Paragon drawing paper with the rough side up.

Draw with HHH pencil horizontal and vertical lines to divide the sheet into one 12×18 rectangle on the left and into two 9×12 rectangles on the right.

In general draw all horizontal lines with the upper edge of the T square blade, and all vertical lines with the left hand edge of a triangle manipulated in combination with the T square.

Draw all lines with the **wedge end** of the **pencil.**

Draw all lines from **left to right** with the pencil slightly inclined to the right. The **principle is to drag** and not to push the pencil.

DRAWING OF DRIVING STUD.

Penciling.

Use HHH pencil for penciling drawings which are to be finished in ink.

Commence at the top and left hand side of a drawing and work to the right and down the drawing.

Draw border lines ½" from the sides of rectangle.

Draw lightly a rectangle enclosing space for title.

Draw lightly a rectangle enclosing space for bill of material.

Draw an unbroken line for center line of stud.

Mark off directly from the full size **scale at one setting (Fig. 4), with** the **wedge end** of the **pencil,** points on the center line through which all

the vertical lines of the stud and attac for the diameter of split pin and ¼" for gr

Draw light vertical lines of indefinite

Mark off on each vertical line in orc points through which the horizontal line in proportionally the horizontal lines for t

The distance across the corners of assumed as twice the diameter of its he middle face as a little less than the diamet

Draw horizontal lines through points.

Draw freehand with light lines, comm sion lines with their included dimensio thread lines, and section lines taken sepa

Inking.

Keep bottle of ink in drawer, not on

Make the width of lines in the printir types of lines used in the drawing correspe

Draw arcs of from 1/16" to ½" radius wi

Fill drawing pens and writing pens perfectly clean the moment they cease t **is an essential.** It should be free from li wrapped around the bottle of Higgins' cloth soaked free from glazing makes a ve

Arcs on top of split pin are tangent bottom is made with a slightly larger rac on stud and washer are made with a radiu

Draw arcs of more than ½" radius witl to the plane of the paper. The compa holding one of its legs, but it should be g at the yoke and turned in a right hand d slight inclination in the direction to wl **Fig. 9.**

The large arc on the end of stud is drawn with a radius of 2″.

Take the radius of the arc directly from the scale by holding the compass parallel with its plane surface, with the needle point shoulder and pen-point just touching its edge.—Fig. 10.

Arcs on nut are drawn as follows: arc on middle face is drawn with a radius equal to the diameter of hole, with its center on center line, and tangent to the bounding line of nut.

Each of the end arcs is drawn through two points: one, at the point of intersection of the middle arc and the bounding line between the middle and end faces; the other, at the point of intersection of an imaginary center line through face and the bounding line of nut.—The radius of arc and its center should be found by inspection.—Fig. 17.

Draw with the ruling pen all the horizontal and vertical visible lines of the object in their order.

Draw 45° chamfer lines on corners of nut.

Draw visible thread lines on end of stud.

Space thread lines about as shown on the plate with a very slight inclination.

Parallel inclined lines are drawn with one straight-edge held fast and another sliding along it at the proper inclination.—Fig. 3.

Draw horizontal, vertical, and inclined invisible lines (this includes thread lines) of object in their order.

Make dashes in invisible lines of object short, of exactly equal length, and almost joining.—Plate I.

Draw center lines, extension lines, and dimension lines in their order.

Make with the writing pen the arrow heads, dimensions, notes, and witness lines. Make the arrow heads narrow and long, not broad and short.

Draw section lines with the 45° triangle placed against T square blade. Space the lines not over $\frac{1}{16}$″ and not less than $\frac{1}{16}$″ apart.

Print identifying letters not less than $\frac{1}{16}$″ in height.

Pencil lines of bill of material.

Ink outside lines, the top horizontal division line, and all the vertical division lines of bill of material to the same width as the visible lines of

object, and the remainder of the horizontal li any auxiliary line, as, center line, extension lin

Print bill of material.

Print title.

Print index marks in lower left hand co identifying letters.

Draw border lines to the same width as the

DRAWING OF WO

Penciling.

Draw border lines.

Block out space for bill of material and title

Draw center line of worm.

Mark off on center line the points through pass.

Draw indefinite lines through points.

Mark off the points on the left vertical line tl lines will pass, including the top and bottom lines the middle of the teeth on each side. The dep pitch.

Mark off points $\frac{1}{2}$″ apart on the horizontal middle of teeth.

Draw 75° lines through the points on the mi that a tooth is opposite a space.

Draw 75° lines by holding fast the T squa and placing the 30° and 60° triangle against th

Draw outside inclined lines of teeth connecti

Draw freehand the broken visible line of th

Draw freehand the extension, dimension, order.

COURSE IN DRAWING.

Inking.

Draw visible and invisible lines of the object (the visible line representing the broken surface should be drawn with ruling pen), center, extension, and dimension lines in their order.

Make arrow heads, dimensions, notes, and witness lines.

Draw hatch lines.

Print identifying letter.

Make bill of material, title, and index marks.

Draw border lines.

DRAWING OF PLUNGER AND CARRIAGE FEED SCREW.

Penciling.

Draw border lines.

Block out space for bill of material and title.

Carriage Feed Screw.—

Draw center line.

Mark off on the center line the points through which all vertical lines in the side view pass.

Draw indefinite vertical lines through points.

Mark off on vertical lines the points through which the dimensioned horizontal lines pass.

Draw dimensioned horizontal lines.

Draw square in end view.

Draw circles in end view.—Take compass radii directly from side view.

Draw arcs of circle bounding square in end view.

Draw remaining horizontal lines of square in side view.

Draw key in side view proportional to size shown on plate.—The arc is less than a semi-circle.

Draw key in end view.—Key is about $\frac{3}{8}''$ wide.

Mark off points on top and bottom line of square threads directly opposite to each other and $\frac{1}{4}''$ apart.

Draw inclined lines through points
a space.

Draw bottom lines of teeth at a dept

Draw diagonal lines and the right h

Draw freehand the remaining arcs a

Draw freehand the extension, dim
order.

Plunger.—

Draw center line.

Mark off on the center line the poir
pass.

Draw vertical lines of indefinite leng

Mark off on vertical lines the poin
pass.

Draw horizontal lines.

Draw top and bottom horizontal li
$\frac{1}{8}''$ from body of plunger.

Mark off on top and bottom center
The center of a coil on one side of p
the middle of the space between two cer

Draw circles of coils with $\frac{1}{4}''$ radii.

Draw inclined lines tangent to c
opposite directions on back of plunge

Draw freehand the arcs and broken l

Draw freehand the extension, dim
order.

Inking

Plunger and Carriage Feed Screw.-

Draw small and large arcs in their c

Draw horizontal, vertical, and inclin

Draw invisible lines.

Draw the diagonals and right hand
Make fine lines.

Draw center, extension, and dimension lines in their order.

Make arrow heads, dimensions, notes, and witness lines.

Draw hatch lines of plunger and adjacent spring oppositely inclined.

Print identifying letters.

Make bill of material, title, and index marks.

Draw border lines.

Submit sheet of drawings to instructor.

Cross-check drawings with another draftsman.—**Every drawing should be thoroughly checked before it is filed as finished.** All mistakes found on the drawing after it has been finally filed should be charged against the checker. The original draftsman should make all changes and corrections.

Check drawings by steps.—

1. Note lines of object.
2. Note dimensions and working notes.
3. Note that the dimensions are to scale.
4. Note arrow heads.
5. Note accents, as, inch marks, foot marks, and degree marks.
6. Note center lines.

DRAWING OF LAYOUT OF PIPING.

·**Plate III.—Layout of Piping.**

Tack to board a sheet of 18 × 24 brown detail paper.

Penciling.

Follow the same general order as instructed for penciling drawings on Plate II.

General hints on penciling lines of object.—

A piping layout is conventionally shown and may be drawn to approximate dimensions.

Commence at the top and left hand and work to the right and down the sheet.

Make piping drawing double the size sho⸍

Draw the tee fitting complete, then the coupling, pipe, globe valve, and so on in orde

Mark off distances on the drawing by doub directly by the compass from the plate.

Inking.

Follow the same general order as instruc Plate II.

The lines of object for "Acid Tank" are : Cross-check drawing.

DRAWING OF 5″ × 30″

Plate IV.—5″ × 30″ Pulley.

Tack to board sheet of 18 × 24 brown detai

Penciling.

Follow the same general order as instru on Plate II.

Hints on penciling lines of object.—

Make drawing to the scale: 6 ins. = 1 ft. to the foot on the full scale).

In left hand view—

Mark off on center line points through hub and rim pass.

Mark off on center line a point midway 1 Draw indefinite vertical lines through poi

Mark off on vertical lines the points thro and the inclined horizontal lines of hub pass of keyway.

Draw horizontal and inclined horizontal lines of hub.

Draw outside circular arcs of rim with a radius six times the width of rim.

Draw arcs of long radii with beam compass. If compass clamps are used, the compass legs should be gripped as close to their points as possible.

Mark off, on end lines passing through rim, points through which the inclined horizontal lines of rim pass.

Draw inclined horizontal lines of rim.

Draw arcs for beads on rim and hub with radii equal to semi-minor axes of respective ellipses in right-hand view.

Draw inclined lines tangent to bead arcs on rim and hub.

In right hand view—

Draw the three circles at rim.—Middle and inside circles represent under edge of rim and bead respectively.

Draw the two circles at hub.—Larger circle represents outside edge of hub.

Draw a light auxiliary circle to represent outside of bead on hub.

Draw light auxiliary lines 60° apart to represent center lines of arms.

Draw ellipses by use of circular arcs. When the major is twice the length of the minor axis, the side arcs of the ellipse can be drawn with radii which are three quarters the length of the major axis. The end arcs should be drawn tangent to the side arcs and pass through the end points of the major axis.—**Fig. 27.**

Draw arm tangent to ellipses and the remaining arms to correspond.

Draw freehand the fillets tangent to bead circles at top and bottom of arms.

Draw arcs of circles at top of arms about to the proportion shown on plate.

Draw keyway in both views.

Inking.

Tear off wrinkled edge and tack to board a sheet of 18×24 tracing cloth with rough side up.

Rub tracing cloth with blotter or a li

Follow the same general order as in

Plate II with the addition that "f" m notes.

Cross-check drawing.

DRAWING OF 1½ P. 76″ P.D. 8″ F BUSHIN

Plate V.—1½ P. 76″ P.D. 8″ Face G

Penciling

Tack to board a sheet of 18×24 bro

Follow the same general order as ins

Hints on penciling lines of objects.-

Make drawing to the scale: 3 ins. = scale where one 3″ length is divided and ing 3″ lengths are marked in feet.

In top view—

Mark off on center line points thro of hub and rim pass.

Draw indefinite lines through points.

Mark off on end lines of hub points of hub will pass, including lines bound on hub.

Draw indefinite vertical lines throug

Mark off on end lines of rim points of rim pass including bead on rim.—Ma

Draw indefinite vertical lines throug

Mark off the width of beads on thei and rim.

Draw inclined horizontal lines connecting corners of beads.

Draw freehand all fillets.

In bottom view—

Draw the three circles at rim.—Middle and inside circles represent under edge of rim and bead respectively.

Draw circles at hub.

Draw a light auxiliary circle to represent outside of bead on hub.

Draw light auxiliary lines 60° apart to represent center lines of arms.

Mark off width of arms and their webs on the bead circles at hub and rim.

Draw inclined lines of the arms and their webs.

Draw freehand all fillets tangent to their respective lines.

Draw section of arm with taper on inside of flange to the proportions shown on plate.

Draw in both views pipe plug and the oil hole under it to the proportion shown on the plate.

Draw oil holes in bushing, locating them in about the position as shown on plate.

Inking.

Tear off wrinkled edge and tack to board a sheet of 18×24 tracing cloth with smooth side up.

Follow the same general order as instructed for inking drawings on Plate II with the addition that "f" marks should be put in with the notes.

Cross-check drawing.

SECOND GRADE.

Make, from sketches of machine parts, finished penciled detail drawings on 18×24 brown detail paper.

Use HHH pencil and 8×10 cross section

An ideal sketch should have the appearan drawing, but all lines should be drawn witho and all notes should be plainly written and no

Sketch objects by this method: First, necessary views. Second, draw in proportio Third, draw all inside and minor lines of dimensions. Fifth, write in all notes.

Refer to Manual of Drawing and learn A 276, 286–296.

Cross-check all drawings.

THIRD GRAI

Make, from sketches of the parts of a sec inked assembly drawing on 18×24 bond pap

Refer to Manual of Drawing and learn A Cross-check drawing.

FOURTH GRA

DRAWING OF ENGINE E(

Plate VI.—Engine Eccentric.

Make finished inked detailed drawing on 1 tric.

Refer to Manual of Drawing and learn A 276, 286–296.

Note.—Separate straps and sheave and draw the two halves of each together.

Cross-check drawing.

DRAWING OF MAIN CONNECTING-ROD FOR TEN WHEEL LOCOMOTIVE.

Plate VII.—Main Connecting-rod for Ten Wheel Locomotive.

Make finished inked detailed assembly drawing on 18×24 bond paper of Connecting-Rod.—**Figure below.**

Refer to Manual of Drawing and learn **Articles: 154, 163, 173–179, 181–186, 188, 190–276.**

Note.—Draw top and side views. Be careful and put in every necessary dimension clearly and do not crowd dimensions. Put in dimensions between center lines of brasses.

Cross-check drawing.

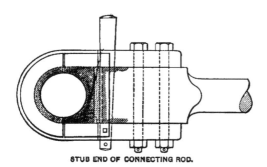

STUB END OF CONNECTING ROD.

DRAWING OF ENGI

Plate VIII.—Engine Crosshead.

Make finished inked detailed drawir head.

Refer to Manual of Drawing and l 276, 286–296.

Note.—See isometric of Crosshead

Cross-check drawing.

FIFTH G

Make finished inked drawing on bc templet for the Air Compressor shown

Refer to Manual of Drawing and l 288–295.

Note.—Make drawing of foundati foundation shown on Plate IX.

Cross-check drawing.

Make from sketch of room a finish of a Floor Plan.

Refer to Manual of Drawing and le

Note.—Make floor plan to type sho

Cross-check drawing.

Make Patent Office drawing from s

Refer to Manual of Drawing and lear

Note.—Make Patent Office drawing

Cross-check drawing.

EXTRA DRAWING.

Make a 12×18 blue print diagram drawing of the 15″ Pillar Shaper omitting Countershaft.

Refer to Manual of Drawing and learn **Article: 306.**

DRAWINGS OF 15″ PILLAR

Plates XII–XIX.—15″ Pillar Shaper.

Make an assembly drawing of the 15″ Pilla of it.—**Figure below.**

Refer to Manual of Drawing and learn **Art**

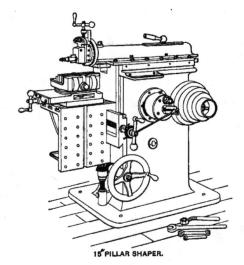

15″ PILLAR SHAPER.

MANUAL OF DRAWING.

MANUAL OF DRAWING.

MATERIALS.

1. The **materials** used in drawing may be divided into two distinct parts. One part would represent the materials on which drawings are made and reproduced before any substances are applied to make that contrast which pronounces it a drawing for the use of the manufacturer or builder in commercial mechanical lines who produces, by means of it, that which has utility and a commercial value. The other part would represent such substances as are necessary to apply to the drawing materials, for the purpose just described.

The materials on which drawings are made and reproduced would include drawing paper, bond paper, tracing paper and cloth, print-paper, etc.

2. The **substances applied** for making drawings would include lead, ink, chemicals for prints, etc.

3. **Drawing Paper.**—The **ideal** drawing paper should be of tough fiber, of uniform thickness and surface, neither repel nor absorb ink before or after it is rubbed with an ink eraser, and take the ink without wrinkling the surface.

4. **Whatman's hand-made paper** approaches most nearly the ideal drawing paper. It is about the most expensive drawing paper made and for that reason it has a very limited use in commercial establishments.

5. **Manilla** wrapping **paper** is a brown paper and one of the cheapest brands of paper made. It should never be used for finished inked drawings.

6. **Keuffel & Esser's Duplex** (or equivalent) drawing **paper** is a brown or cream colored drawing paper and is a little better quality than the average commercial grade. It will bear a fair an and inking.

7. **Egg-shell paper** with a linen back and ro ally used for a drawing made to stand out in effect. It is a very durable paper and will bear

8. So-called **indestructible paper cloth** is made is very durable and will bear rough handling.

9. **Bond paper** is a thin and comparatively which has the distinctive advantage that the c directly printed from on print paper or cloth, an omission and commission made in tracing from a

Bond paper is very easily wrinkled and can quickly. Therefore it requires most careful h drawing on bond paper should not, as a rule, be except a fine **ink-eraser**.

10. **Cross-section paper**, whose checks are c the advantages found in bond paper and the add on it without the assistance of a scale.

11. **Drawing paper**, when it comes in rolls, with the convex surface next the drawing boa should always be on top.

12. **Drawing paper** is occasionally **stretched** not to be taken off for some days; but after it is ally contracts and the scaling is affected.

13. The paper is stretched on the board in

first, clip off the corners so that the four edges can be folded over for three fourths of an inch; second, wet all the surface, except the turned edges, with a sponge; third, lay the wetted surface next to the drawing board; fourth, smooth the paper by rubbing from the center out to the edges; fifth, glue down the edges with very strong gum arabic or like material.

14. A paper **drawing** may be longer **preserved** by **mounting** on straw board and varnishing its surface with white shellac.

15. **Tracing Paper.**—Tracing paper is a firm and transparent paper having a smooth and glossy or oily surface and can be easily printed from on print paper. It should be of tough fiber, uniform thickness and surface, neither repel nor absorb ink before nor after it is rubbed with an ink eraser, and take ink without wrinkling the surface.

Tracing paper should not generally be used when permanency of a drawing is the chief object.

In order to facilitate a design, tracing paper is used sometimes for **tracing alternate positions** of a piece of mechanism, when the relation of the piece to the other parts of the machine is known.

16. **Tracing Cloth.**—Tracing cloth or linen should have one of its surfaces well glazed and no open pores. It possesses the same properties and uses as tracing paper, besides being more permanent. It has a more extensive use commercially.

Before tracing cloth is laid on the board the **wrinkled portion** along the edges should be **torn off** and its surface should be stretched smooth. (Tracing cloth will tear straight parallel to its edges only.)

17. There is considerable discussion and contention as to whether the glazed or unglazed side of tracing cloth should be used. There are **advantages** and **disadvantages** in using either. It must be admitted that the glazed side was primarily intended for use; that drawing ink, especially red ink, will eat deeper into the unglazed surface with consequent difficulty in rubbing; that it is usually rolled with the glazed side in, which would naturally bring the glazed side on top, as the convex surface is placed next the board; that the tracing does not curl so much when inked on the unglazed side as on the glazed side; and also

that the tracing will eventually smoo[...] side if placed in the drawer with th[...] writer must insist that, from his pr[...] no change from ancient custom shou[...] ciled on the cloth, when it is absolu[...] should be used in order to see the p[...]

18. Before using tracing cloth a[...] exposed to the air, it should be **rubb**[...] then brushed off thoroughly.

19. Tracing cloth is very **suscep**[...] and it will become taut or loose with[...]

Water will **ruin tracing cloth** an[...] tion from the hands is prevented fro[...]

The sizing can be dissolved or s[...] very **desirable pen wiper.**

20. **Tracings** can be **cleaned** w[...] highly volatile substance.

21. **Sheet Celluloid.**—Sheet cell[...] tracing paper for tracing alternate p[...] is more durable than either tracing[...] easily.

22. **Print Paper.**—Print paper[...] considerable handling, and should[...] ordinary sunlight.

23. To print from a drawing pr[...] the printing frame with the ink lin[...] tized surface of the print paper ne[...] time to the sunlight and then remov[...] tion for a suitable time, which will v[...]

24. If there are **certain lines, fig**[...] **desired** on the blue print, they can l[...] of the print by placing a piece of **opa**[...]

25. The **principle** involved in [...] after the sensitized surface of the [...]

light and passed through a fixing solution. That part of the surface of the print paper under the inked lines, figures, etc., on the drawing is shielded more or less (depending on the color of ink used) from the light; therefore after the fixing solution is used, the necessary contrast, which is desirable on the print, obtains. It is thus evident from the known relation of colors to light, that, when the most distinct lines are required on the print, the blackest and most opaque ink must be used to entirely exclude light from the surface of the print paper.

26. **Colored inks,** which are of course **not so impervious** to the light as black ink, are occasionally designedly used on drawings to give a less distinct line on the print; but, for commercial use, thin black lines of different character from the lines of projection on the drawing are far more desirable for many reasons which are noted through the text, and serve the same purposes that colored lines usually do in commercial mechanical drawings.

27. In a *resumé* of the foregoing it is evident that the time of exposure of the print paper to sunlight varies according to the sensitiveness of the chemicals used; with the material upon which the drawing is made; with the substances applied in making the lines, figures, etc., on the drawing; and with the intensity of sunlight (which is usually more effective in Winter than in Summer) or artificial light.

28. Apparatus of several designs have been devised in which the **electric light** is used **for printing** and they are especially convenient in cloudy or stormy weather.

29. All **print paper** should evidently be as **fresh** as possible; and, when not in the printing frame, it should be kept from the light in a covered can or case placed in a dark room.

30. It must be borne in mind that print paper usually un-uniformly **shrinks,** and, therefore, it must *never* be scaled for actual dimensions.

31. **Blue print paper** is decidedly the most used commercially. It produces a white line on a blue field.

32 The field of a blue print darkens according to the amount of exposure, and, as none of the black drawing inks are absolutely opaque, the value of a print can be annulled by **over exposure.**

33. After exposure to sunlight, **blue print** paper is in **water** (which is the fixing solution), and then the printed surface should be washed off with a under pressure.   The print should remain in the minutes.

34. The ordinary blue print paper requires abou **exposure** to bright sunlight.

35. There is a **quick blue print paper** which re utes' or less exposure in bright sunlight when pri made on tracing paper or cloth or their equivale

36. In **printing from bond paper and cross s** exposure to the light is required than for printing tracing cloth.

37. A **good print** shows everything distinctly of the field.   A **print with a light field** may be n requires less time to print, and additions and cor with black drawing ink.

38. **Blue print cloth** is printed in the same mann It is much more permanent than blue print paper usage.

39. **Black print paper** produces a black line c immersed in a chemical bath first, if there is no the coating of the paper.   It is afterward carefully same manner as blue print paper.

40. **Brown print paper** produces a white line should be immersed in a fixing solution and then v same manner as the other print papers.

41. **White prints** or " **Vandykes** " may be made brown print paper from a negative made on bro negative is made by placing the drawing in the ink lines next the sensitized surface of the brown p tive white print is then made by placing the ne frame with the print next the sensitized surface paper.

42. A print may be longer preserved by **mounting on straw board** and varnishing the surface with white shellac.

43. **Inks.**—Black drawing ink should be opaque, waterproof and non-decomposable; and should flow freely, dry quickly, and not eat into the surface of the drawing material.

44. Black drawing ink can be **prepared from stick** India ink, but it is more convenient when purchased in the bottles of the prepared commercial waterproof drawing ink.

45. The commercial prepared waterproof drawing ink must **never be thinned.** If the ink does not flow satisfactorily, examine the pen for the source of the trouble, or if the ink has actually changed (which is almost invariably not the case), procure another bottle.

46. **Red** or carmine **drawing ink** (used only when absolutely necessary) should be waterproof and non-decomposable. It should flow freely, dry quickly, and not eat into the surface of the drawing material.

47. All ink, when in use, should be **kept in** an open **drawer,** within convenient reach, or at a sufficient distance away from the drawing to prevent it from being upset on the drawing.

48. When the **fibers** of a surface have been **torn up** by careless rubbing, the prepared varnishes painted on it or the **rubbing** of the affected surface **with soapstone,** hard beeswax, bone, or the end of the finger nail, will effectually **prepare** it for **inking.**

49. **Substances used in Preparing and Altering Blue Prints.**—The **sensitizing chemicals** for **blue prints** are prepared from several formulæ in varied proportions.

A good sensitized surface is prepar a solution by weight of

Citrate of Iron and Ammonia
Water. . . . .

Then make a solution of

Red Prussiate of Potash. . . . .
Water. . . . .

Mix equal parts of the two solution: two minutes to a paper having a h off the superfluous liquid and hang u will have a bright yellow hue.

The sensitizing **solutions** can be l when mixed must be **kept from** the li

50. Soda, Potash, Quick-lime, or and a little gum arabic added to kee paper, will produce a white mark on th tions are **used for making alterations** c

Chinese white and other commerci alterations, but the writer believes tha for the purpose of **making correction** **use black ink** on a **light print and red**

INSTRUMENTS.

51. The proper **selection of instruments** is of prime importance. It is universally conceded by first-class draftsmen that good instruments are absolutely essential for the best execution of drawings in the shortest time.

The term **"good instruments"** does not **necessarily** imply that a draftsman should have all the new-fangled specifics that are afloat on the market; as, **section liners, dotters,** etc. A *good instrument*, legitimately interpreted, is one which is indispensable and of the best grade.

52. **Drawing Board.**—The drawing board should be made of soft and well-seasoned wood of uniform grain; should have two adjacent edges straight and at right angles to each other; should have its working surface very slightly crowning in the center; and should be designed to allow for the changes due to atmospheric conditions.

53. If a permanent **working straight-edge** on the board is desired, a heavy cast-iron or steel strip may be securely fastened to it.

54. The **truth** of the **straightness** of the working edge of the board can be tested by applying a standard straight-edge.

55. The **convexity** or planeness of the **top surface** of the drawing board can be proved by applying a standard straight-edge or the edge of a T square blade to its surface.

56. To **prove** the **squareness** of the **working edge** of the drawing board to its lower edge (**Fig. 1**) proceed as follows: first, place T square head against working edge of board and draw two parallel lines, one about one inch from the upper edge of board, and the other about one inch from the lower edge of board; second, with the T square head placed against the working edge of board, mark off on the T square blade a point whose distance from the edge opposite the working edge of board will be about the length of the T square head; third, with the T square head placed against the working edge of board, sight and mark off from the point on the T square blade a corresponding point on each of the

parallel lines; fourth, place the T square head board and draw a line through the point mar: allel lines and note if it passes through the c other line. Absolute squareness is unnecessary

57. **T Square.**—A T square of the best gra a fixed head and blade, with ebony lined edges

The **blade** should be dovetailed or let into surface of the head should always be flush w. the drawing board.

If the **blade** is **fixed** to the head with screw: should be used.

58. As a rule, **adjustable heads** are undesir a double head is practicable and is recommend manently fixed, as specified above, and the otl and fastened with two binders.

59. The **blade** should have its working edg sixteenth of an inch thick, and the ebony lining

60. Experience has proved that the cellulo: some slight advantages, warps and loosens, w isfactory.

61. To **prove** the **straightness of the edge** o ceed as follows: first, draw with it a line; se and place the edge of the blade to exactly coir line which was drawn; third, draw another coincidence of the lines which were drawn, 1 straightness of the edge.

62. The truth of the **straightness of the he** ing a standard straight edge.

63. A good **pear-wood T square**, designed and altogether decent instrument.

64. A **nickel plated** steel **T square** has the decided advantage of maintaining its truth; but its surface gathers dirt and smuts the drawing, which fact makes it less desirable than the other types.

65. **Parallel Straight-edge.**—A parallel straight-edge is fastened to a cord or wire running in grooved rolls which are secured to the corners of a drawing board, replaces the T square, and, consequently, the working edge on the drawing board.

There are other arrangements of the parallel straight-edge and nearly all of them are most desirable.

66. **Triangles.**—The **transparent** triangle appears to be the most popular triangle used at present. It is subject to change in planeness and accuracy and the very best material used in its manufacture is none too good. It possesses the advantages of not obstructing the view of anything on the drawing when in use, and of keeping a drawing cleaner than other triangles.

67. The **hard rubber** triangle warps somewhat, but usually keeps its planeness longer than the transparent one, and smuts the drawing more.

68. The **pear-wood** triangle is cheaper, but subject to change due to atmospheric conditions. It does not smut the drawing like rubber.

69. The **nickel plated steel** triangle maintains its planeness and truth, but, like other steel instruments, it gathers dirt and smuts the drawing.

70. The **30°✕60°** and **45°** triangles are most commonly used, but there are a large variety whose sides make other angles with each other.

71. If the central portion of a triangle is cut out, the **inside edge** may be **beveled** on one side to facilitate handling when it is transferred from one part of the drawing to another.

72. An **inserted knob** is sometimes used and perhaps is more convenient than the beveled edge for picking up the triangle.

73. Triangles should be of **uniform thickness** and their **outside edges** should **not** be **recessed** to avoid the danger of ink being drawn from the ruling pen upon the drawing.—The pen should be free from ink on the

outside and always held with its axis in of the paper.

74. The **combination triangle** desig Conn., combines all the functions of angles, except the drawing of parallel other, and a line making 75° with th knob and generally takes the place of used in combination with the T square

75. By properly combining the T lines can be drawn **making 30°, 60°, vertical.**

76. By properly combining the T be drawn **making 45°** and **90° with th**

77. By properly combining the T triangle, **lines** can be drawn **making 1 vertical.—Figs. 1, 2.**

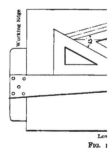

Fig. 1

78. By properly combining a trian and 45° triangles, **parallel** and **perpend**

79. Both triangles are **proved for t** lows: first, place the short side of th second, draw a line; third, reverse th

directly over the first; fourth, observe the coincidence of the lines, which will prove that the angles are correct.

80. The **30°** and **60° angles** can be **proved** for their truth as follows: first, place the short side of the triangle against the edge of the T square

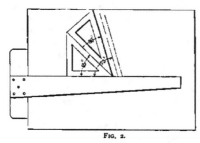

FIG. 2.

blade; second, draw a line through a point with the 60° side of the triangle; third, turn the triangle over and with same side of triangle draw another line through the same point; fourth, draw a horizontal line across the extremities of the lines just made; fifth, measure with

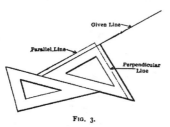

FIG. 3.

dividers the inclined lines to prove the equilateral triangle which obtains for a true 60° angle.

81. The **45° angle** can be **proved** for its truth as follows: first, place

one of the short sides of the triangle against blade; second, draw a line through a point angle; third, place the other short side of th of the T square blade and draw another lir with the 45° side of the triangle; fourth, dr. the extremities of the lines just made; fifth, inclined lines to prove the isosceles triangle w angle. (This should follow the proof of the o

82. The triangle should always be **used** to the **light** to avoid shadows.

83. The **working edge** for a right hand left; and the drawing table should be arra to have the light on the proper side of the tria

84. **Northern exposure** and diffused ligh is undoubtedly the ideal light **for a drafting ro**

85. **Scale.**—The scale is made in many forr **lar boxwood scale** is the most popular. It is ¢ and will admit of combining, in one scale, all usually used in making any particular type drawing.

86. A **scale guard** should be used with th to obviate the usual experience of inconvenie on the wrong side when it is in use.

Undoubtedly the **nickel plated sheet-met** most accurate work, which is vitally essential in

87. The **marking off** of **distances** is gre: the pricker point or pencil point down the inde indicate the graduations.—**Fig. 4.**

As there are but two graduated edges on th of frequently **turning the scale** is eliminated.

The ancient cry, that the eyes are injurec from the nickel plated steel scale, has been un sees no reason why a steel scale should injure enameled one.

88. A **flat boxwood scale** with beveled edges has less pitch on its sides, and, for that reason, is considered by some to be more quickly and easily

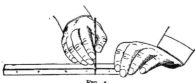

Fig. 4.

read than any other form. It is not, however, as readily manipulated as any of the others.

89. Both sheet steel and boxwood **scales** can be obtained **in sets** to suit the requirements, but the former scale is recommended as superior to all others.

The ordinary **scale** was **never designed for a straight-edge** or ruler, or for any other purpose than to take distances from.

90. When a **drawing** is made **on reduced scale,** the full size scale *should not* be used, but the particular scale which is arranged and adapted for the purpose. For example: a drawing may be made one fourth size in English measure, or its equivalent, three inches to the foot, which indicates that a graduation is to be found where every three inches in length is marked in feet; and also, one three inch length is subdivided into twelve equal parts which are marked inches.

Such **terms** as ¼ scale or size, and ¼ inch to the foot must not be confounded when the proper graduations are being sought on the scale. A moment's reflection will make it clear that the former stands for three inches to the foot and the latter for one fourth of an inch to the foot; and their respective scales will be graduated as indicated above.

91. A **scale** can be **proved** for accuracy as follows: mark its equal subdivisions off on a straight line and then reverse the scale and note if the equal subdivisions coincide.

92. The **limit of accuracy** in scaling should be, in commercial mechanical drawing, not more than one-hundredth of an inch.

93. If any number of equal or un off on a right line, they should alway **scale.**—Fig. 4.

When several distances within the li there is no line on which to lay the sca **against** the scale.

Consecutive distances within the li at **one setting.**

94. **Protractor.**—The protractor : angles which cannot be gotten from th

A **steel protractor** should be of f as little surface as possible to come in

Cheap German silver and celluloid

95. **Universal Drafting Machine.** is a device which is advertised to T square, triangles, protractor and sca for it and it is well recommended by p

96. **Curved Ruler.**—The curved r certain irregular lines, but it should n practically substituted.

A set of curves is often selected fo **universal curved ruler** will often me average logarithmic spiral curve is rec

Curved rulers are usually made of of steel, as are found in triangles; anc advantages and disadvantages, as i unquestionably, a transparent curve other.

97. **Drawing Pencil.**—A good dr and conventional) is essential for proc lines.

The pencil should be made of a ι out; and a suitable grade of hardnes drawing which is to be made.

A **soft pencil** draws **smoother,** easier and faster than a hard pencil.

A drawing which is to be either inked or traced over should be made with a soft pencil of **HHH** or **HHHH grade.** A drawing made for permanent use and not to be inked or traced, should be made with a **HHHHHH** to **HHHHHHHH** grade pencil.

98. The **form** of the wood, which encases the lead, should not be circular in section.

Sharpen both ends of both pencils, one end of each to a cone whose apex is slightly blunted, and the other to a wedge whose sides are flattened ⅜″ from the point and whose edge is slightly elliptical but keen. The "lead" should project from the wood not less than ⅜″ and not more than ½″.—**Fig. 5.**

99. The flat **point** should be **used** for drawing lines and the conical point for free hand work.

When lines are drawn with the **flat point** of the pencil, it should be used for **marking distances** from the scale.—**Fig. 4.**

The **pencil, when in use,** should have all of the available surface of the lead in contact with the straight edge which guides it; and it should be held slightly inclined in the direction to which it is being drawn.

The pencil should be pressed lightly down and drawn from left to *right* and away from the manipulator.

100. The **Artist's Pencil** with movable lead is a very desirable substitute for the ordinary drawing pencil. Two of them should be procured for convenience; one with a conical pointed and the other with a chisel pointed lead in it.

101. **Ruling Pen.**—The ruling pen is about the most important instrument found in what are known as the case instruments.

It is of **prime importance** that the **greatest care** should be exercised in the **selection of the pen,** and it is of more importance to keep the pen in first class working condition.

Wedge Point

Conical Point'
FIG. 5.

Poor execution is often due to poor material negligence in repairing it.

The ruling pen should have its blades mad steel. It can be **tested** for the softness or hard drawing a Swiss file lightly across its nibs.

The insides of the **blades** of the ruling pen sh as possible.

One blade should be arranged with a prope device **for opening** the **blade** quickly and widely.

Many draftsmen oppose the **hinged ruling p** the joint for the hinge impairs a certain rigidity the pen is in use. It is undeniably true that a b not jointed, but hinged ruling pens have been some few on the market, which meet all the es rigidity. It is vitally necessary that the hinged with an effective device for taking up the wear at

102. A **pen must be kept** thoroughly **clean,** ? sarily thoroughly cleaned by drawing a rag throu close together, but, if it is necessary, each nib sh smoothly polished. Hence, the convenience g: sense wide-opening pen.

103. The **handle** should be made of a mat broken. Bone or ivory handles are undesirable broken, but an aluminum or ebony handle is ve larly the former.

104. The ends of the **nibs** must be kept in pr condition obtains when the ends of the nibs are are not over one thirty-second of an inch; are knife; and both touch the paper when the pen is a plane perpendicular to the plane of the paper.

105. To **repair** the ends of the **nibs,** proceed first, clean the pen thoroughly; second, hold the p piece of white paper and close the nibs until th third, hold the pen with its axis perpendicular to

oil stone and move the stone over the points of the nibs until an arc of not over one thirty-second of an inch radius is rounded on the points (**Fig. 6**); fourth, open the blades as wide as possible; fifth, hold the axis of the pen at right angles with a line along the length of the stone and,

108. When a line is being dra... axis in a plane perpendicular to the... in the direction to which the pen is l... on the pen, drawn from left to right a...

FIG. 6.   FIG. 7.

Fic.

with one blade at an angle of about 30° with its plane surface (**Fig. 7**), move the point of the nib about one-half inch back and forth along the stone while the pen is being slightly oscillated about its axis to maintain the roundness on the back and grind the point of the nib to an edge as sharp as a pocket pen-knife. — This operation should be performed on each nib separately and with extreme care to avoid grinding off the end of a nib, thus making one nib shorter than the other; sixth, rub lightly, with the thin edge of the stone, the inside surface of each nib to remove any possible burr; seventh, test the pen thus: clean nibs thoroughly; close the nibs until they are about to touch; fill with ink to a depth of not over three sixteenths of an inch; hold the pen with its axis slightly inclined to the right and in a plane perpendicular to the plane of the paper and draw it lightly from left to right over the paper until the ink is exhausted.—**Fig. 8.** The pen should not fail to make a line while it is moving across the paper and the line should be clean cut.

106. The pen should always be **set** the proper width by the eye, which is done by holding it to the light or over a piece of white paper.

107. The pen should ordinarily be **filled** not deeper than three sixteenths of an inch unless an extra wide line is to be drawn.

If the ink does not flow, do not ruin the pen by jabbing it into a piece of wood or paper, but start the flow by drawing it across the little finger or a wet sponge; then if the ink fails to flow, wipe out and refill the pen.

When a pen is in use, care should... of the **nibs** and that the guiding strai... the inked line. Carelessness and ... be followed by ink flowing under th...

Since most drawing inks dry q... always be **wiped thoroughly** after us... ruption.

It behooves every draftsman to ... Theo. Alteneder & Sons, Philadelp... discussion of the ruling pen and oth...

109. A medium size **pen** can b... **widths**; and the belief held by mar... are necessarily drawn with a small a...

110. If red ink must be used, s... pen should be **reserved** for it.

111. **Compass.**—The best compa... ening bar, needle point leg and pen... of German silver. They should be ... ing when ordinarily handled, and ... accidentally dropped.

All **joints** should have large, ... surfaces and simple and well-design...

The compass should have simple and effective **means for holding detachable parts.** Undoubtedly the round shank and corresponding split socket with its clamping screw is the best design obtainable for connecting and disconnecting the detachable parts of the compass legs.

The cylindrical **handle,** on the yoke which straddles the joint at the top of the legs, is indispensable for facilitating the manipulation of the compass.—**Fig. 9.**

The pen, pencil, and needle point legs should be provided with **flexible joints,** designed according to the specifications given for corresponding joints in the compass legs; and also, the joints should be as near to the working points as practical.

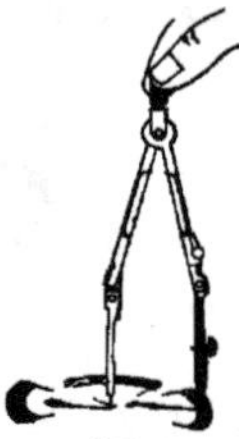

FIG. 9.

The **pen** should be designed and cared for in the same manner as specified above for the ruling pen.

The **pencil** should be of a material described above in "Substances Applied, etc." and of nothing softer than a HHHHHH grade. It should be so sharpened that the point has a very flat elliptical section.

112. The **needle points** should be made of the best tempered steel with one end conical, and the other end with a very short, sharp, and fine point, terminated by a square shoulder.

The needle point and lead should be a good fit **in** a split socket and held in place with a clamping screw. The **device** for **securing a needle point** by the point of a screw should be shunned.

A **needle point** should **not be screwed** into the socket.

113. The **shoulder end** of the needle point **should** always **be used** when drawing circular arcs, and should be adjusted so that both the shoulder and pencil or pen points touch the drawing, when they are perpendicular to the plane of the paper.—**Fig. 9.**

The **needle point** should be occasionally **ground** on the oil stone until it is sharp enough to enter the drawing paper when a pressure due solely to the weight of the compass is applied.

114. If a screw in the compass or bow instruments works hard or becomes rusty, a little **graphite** rubbed off the dr cate it.

115. When a scale distance is used by the com by it directly from the scale, which is accomplish pass level with the surface of the scale and apply the shoulder of the needle point and the end of just touch the edge of the scale.—**Fig. 10.**

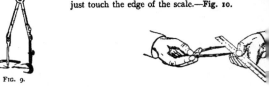

FIG. 10.

116. When a large **circle** is being **drawn wi** inserted, the compass should *never* be steadied of the legs; but it should be gripped, as before, b at the yoke, and very slightly inclined in the direc pass is moving, which direction should always be wise.

117. **Beam Compass.**—The beam compass is circular arcs.

118. The **trammel points** and their attachme materials used for similar purposes in the compa the same manner as prescribed for the regular com

The needle point leg should be fixed; the pen be interchangeable; and the compass should be pr ter adjustment **attachment.**

The **beam** should be designed for stiffness an running of the trammels along its length.

The trammels should be provided with a stro device.

119. The metal **tubular beam** with a split socket on the trammels for binding is recommended.

When an arc of long radius is drawn a spring wheel attachment should be used close up to the pencil or pen point.

120. If a set of good trammel points cannot be had, it is recommended that the **beam compass clamps** listed in the "Supplies required for the Course in Drawing" be procured. The clamps make use of the ordinary case instrument attachments and can be procured from the Cornell Co-op store.

121. **Dividers.**—The dividers should be made of materials which are used for similar purposes in the compass, and should be cared for after the same manner as prescribed for the regular compass.

One **leg** should be without a joint, and the other leg should be provided with only a hair-spring adjustment.

The **needle points** should be conical.

122. If an **exact length** is to be **laid off** with the dividers, a large multiple of that length should be laid off first directly with the scale on a right line, and then exactly sub-divided into the desired exact length by the dividers.

123. **Bow Instruments.**—The bow instruments consist of the spacers, pencil, and pen. They should be made of the best tempered steel, with a metal handle and a center adjusting screw.

The spacers, pencil, and pen are used and cared for in the same manner as prescribed for the dividers and compass, but they are distinctly **designed for short lines and arcs** of short radii.

The so-called "self adjusting needle point" or **"drop" bow pen** has the particular advantage of requiring no adjustment of a needle point to suit the length of the pen or pencil but the needle point end should be made with a shoulder and not conical as is usually done.

124. **Pricker Point.**—The pricker point is used for marking off distance more accurately from the scale than can be done with the drawing pencil.

A fine **sewing needle**, with the eye end firmly driven into a non-cylindrical handle, **makes** a desirable **pricker point.**

A ruling pen, with the **pricker** [
purchased for a few cents extra.

125. **Case for Instruments.**—A
pen, compass and attachments, divid

The folding **pocket case** is undo
good case can be made of chamois
the chamois skin would be incidentall
which should be done every time the

126. **Writing Pen and Staff.**—T
and important in a draftsman's outfi

The pen staff should have a **bulgi**
and should be gripped close to the pe

The pen should be comparatively
ink.

"**Crow quill**" and "**Falcon**" pen
mercial mechanical draftsman's outfi

Care and common sense must l
writing **pen** for the different grades

No more **pressure** should be app
ruling pen, and the erroneous beli
worked under pressure, is the prop
pencil point in the mouth.

127. A good **wiping rag** is an es
A good one is usually found wrap
pared drawing ink.

Tracing cloth when soaked free
sirable **wiper.**

128. **Pencil Pointer.**—A good p
or a sandpaper pad.

There are several good devices
which should be selected with scrutin

A **pocket-knife should never be**
drawing pencil.

129. **Pencil Eraser.**—A pencil e

not the usual velvet type. It should clean the surface of a drawing with very little rubbing and leave it perfectly smooth.

130. A **sponge rubber** is found convenient for cleaning the surface of a drawing without obliterating the pencil lines.

Fine **crumbs from** stale **bread** effectually clean a drawing.

131. **Ink Eraser.**—An ink eraser should be made of a fine abrasive substance which is quick cutting and non-heating; it should leave the surface of the drawing perfectly smooth.

Fine pulverized **pumice stone** will very effectually take out ink lines when rubbed on a drawing.

132. A **steel eraser** is indispensable in the hands of a skillful drafts-man and a pocket knife is no proper substitute for it. It should be made of the best tempered steel with a wooden handle and its edge must be kept keen.

When a steel eraser is used, the ink should be just taken off the drawing and no more.

It is much better for the unskilled person to **take off** only a **part of the ink with the steel eraser and finish with the ink eraser.**

When **lines** are erased from a tracing they should be **thoroughly rubbed** off to avoid vagueness on the print as to what was intended on the draw-ing.

133. **Eraser Shield.**—An eraser shield is sometimes found useful where only a particular and very small part of the drawing is to be rubbed.

An eraser shield can be made by slitting a piece of **sheet tin,** brass, or celluloid.

The straight-edge of a piece of brown dra' very well **for a shield.**

134. **Oil Stone.**—An Arkansas knife oil ston(in a draftsman's kit and should be used when ne index of a good workman when his tools are sharp

135. **Tacks.**—The tacks for holding dowr small heads.

The 1 oz. **copper tack** is recommended; b tack is a nuisance. Its head breaks easily and drawing board to be picked out before the b(as is often done when it is repaired.

Thumb tacks should have a small flat head r a sharp point.

136. **Tack Lifter.**—The tack lifter is a conve It will, doubtless, preserve the original shape o pocket knife, which is so often resorted to as a $\frac{1}{4}''$ or $\frac{5}{16}''$ steel wire, three or four inches long, ' an edge and bent into a claw shape, will serve as

137. **Machinist Scale and Caliper.**—A s and a six inch reversible inside and outside cali venient in taking measurements when a draftsm a piece of mechanism that is already made.

138. **Folding Rule.**—A six foot folding rule to have in many ways. It is often a suitabl(caliper for getting the approximate dimensions (which is being sketched.

TECHNIC.

Commercial Mechanical Drawing.—A commercial mechanical draw-ing is fully defined when such conventions are present as to enable the manufacturer or builder in mechanical lines to erect his buildings; to select and arrange his equipment for most economical production; and to produce, by means of the drawing, and with the greatest facility, that which has utility and a commercial value.

Commercial mechanical drawings can be divided into **three general types** which may be termed: **General Plans, Machine Drawings,** and **Patent Office Drawings.**

139. **General Plans** would include such drawings as: **lay-out drawings, foundation drawings, piping drawings,** etc.

140. **Machine Drawings** would include: **working sketches, scheming sheets, detail drawings, assembly drawings,** and **diagram drawings** of machines.

141. **Patent Office Drawings** are in a class by themselves and are essentially such drawings as are required by the United States Patent Office when a claim for a patent on a mechanical invention is presented.

142. **General Plans.**—Drawing plans of buildings intended for mechanical purposes are often made with nothing more on them than a section of the walls; with the doors, windows and supporting piers for columns located; and sometimes a very general arrangement of the inside equipment shown in part or the whole.—**Plate XI.**

143. **Layout Drawing.**—A layout drawing is often a plan of a floor of a building with the positions of different parts of the equipment located to scale. It is also, sometimes, a preliminary drawing of a part or a whole of a machine and serves the same purpose as the scheming sheet which is defined below. The former sense, however, will be accepted in the Manual.

In **locating** a large number of **machines** or other equipment in a plant, they should be drawn first to scale on a sheet of fairly stiff drawing paper,

cut out, and then moved on the pro factorily arranged; after which, they

144. **Piping Drawing.**—A piping the piping separately from other equip

If there are **several kinds of piping** pipes, water service pipes, etc., and they should be all drawn on one shee each class by itself can be drawn on piper.

In laying out piping drawings, r elaborate views of the fittings, but si **represent them.—Plate III.** All v connections, must be plainly indicate

Do not locate pipes too closely t the piper.

Always **show an outline of the w** pipes run and with which they are c

145. **Foundation Drawings.**—Fc nature of the foundation upon whi whether it is in wood, stone, concret anchor bolts, with their attached cap and also the main outlines of the mach

146. A **templet drawing** usually which are nailed together, and with l relative dimensions between them that bolts in the foundation.—**Plate IX.**

If **brick work** is shown in a found₂ the corbeling should show the outline

147. Drawings of **boiler settings** of Foundation Drawings.

148. **Machine Drawings.**—When a new machine or any of its parts has been conceived, the next step is to develop it on drawings.

Naturally the most **simple drawings** will be **made first** and will be followed in logical sequence by more complete drawings.

A machine drawing must be made simple, clear and direct. All attempts to imitate the product of the camera or to display impractical rudimentary knowledge is foreign to the practice in the United States at present, and must be eliminated in order to harmonize with the existing state of affairs.

To make a machine drawing properly and most intelligently, the draftsman has no other recourse than to be a thorough mechanic himself, which implies a working knowledge of pattern making, molding, forging, and machining.

149. **Sketches.**—Preliminary freehand sketches should be made with a soft black pencil and only the essential outlines of the proposed piece are needed.

Several sketches may be required before one is satisfactory; and all **sketches** made should invariably be **saved** for possible future reference.

It is a rule, frequently followed by designers and inventors, to make **several sketches of the same thing** and it often happens that they return to and use the first sketch made.

150. **Cross-section paper** is invaluable **for sketching** as proportion is secured without the use of the scale.

151. **Copying-pencil sketches** are frequently made to serve the purpose of a small drawing or blue print. They should be made with a copying pencil on checked paper and duplicated by means of a damp cloth and a letter press.

152. **Scheming Sheet.**—A scheming sheet is made up of the accepted freehand sketches drawn to scale and with somewhat more detail shown.

They should be **drawn in with a HHH** or HHHH grade **drawing pencil** on drawing paper and no elaboration to completeness is usually needed.

153. **Detail Drawings.**—Detail drawings are undoubtedly the most important, and, surely, the most extensively used drawings.

The details in the execution of a properly m complete history and exposition of the art of M fore, what has been written, in the paragraph d "Machine Drawings," distinctly applies to de be strictly adhered to.

After the parts on the scheming sheet have drawing, they are then completely detailed for of the shop. Consequently, a detailed drawing **ing**, and, as such, it is his official order for doin should be held responsible only for failure to Anglo-Saxon notes on the drawings.

The **projections** and conventions are secor illustrations to interpret what the figures and no

The **number of parts** of a machine **detailed or** on the methods used in doing the work; or upor are to work from the drawing; or upon both.

Detail drawings are usually pencilled on lig and traced on tracing cloth or pencilled and ir printed on blue print paper in the order given.

For convenience of analysis, the consideratio involved in the complete execution of a detail c under separate headings.

154. **Types of Detail Drawings.**—The num detailed on a drawing should be according to th the methods of doing it. If a shop is manufa where six or more are built exactly alike and a many parts would be detailed on one drawing distributed among a larger number of men.

It is the practice with some firms and the others to make **individual** or separate **drawings** parts.

It has been the custom with some firms to **and forging details** on **separate** drawings and particularly advantageous to firms which build.

When a single new machine, or a section of it, is built for the first time, it is often advisable to show the **parts in partial assembly** and fully detailed on the drawing. Such an arrangement assists the mechanic to make his allowance for proper fits and adjustments. This type of drawing might serve as an assembly or general drawing, and, perhaps, it is more properly called an assembly detail drawing. A bill of material should accompany an assembly detail drawing.—**Plate VIII.**

155. **Sizes of Drawings.**—The size of the detail drawing is dependent on the type of drawing, the size or sizes of the part or parts detailed, and the scale to which they are drawn.

The following **sizes** are good average ones, as they can be **cut very economically** from the commercial widths of drawing paper: 6×9, 9×12, 12×18, 18×24, 24×36, 36×48 and 48×72.

156. **Border Lines.**—Border lines are of no material assistance to the mechanic or to any one else; except, possibly, to the draftsman who may comfort himself with a doubtful fact that the appearance of the drawing is enhanced.

The writer believes it to be the growing tendency to **leave off all border lines,** but, when they are used, they should not be over one thirty-second of an inch wide and have no fancy or elaborate corners.

157. The **border line** should be placed **one half inch from** the **edges** of the sheet; is the first thing to be drawn in a pencil drawing; and the last thing in an inked drawing.

158. **Match Lines.**—Short lines, drawn with the T square on each side of the drawing, are convenient for resetting a drawing after it has been removed from or accidentally shifted on the drawing board.

159. **Titles.**—A title is an essential part of any drawing and a particular and invariable place on the drawing must always be provided for it —**Plates I-V.**

A **place** at the lower right hand corner of the drawing should be blocked out for the **title** directly after the match lines are drawn.

The **title should comprehend:** first, the type of drawing, as, a sketch, assembly, or detail; second, the name of, or the part of the machine

drawn, or both; third, the name of
fourth, the address of the firm; fifth,
ing; sixth, the name or initials of the
sible for the finished drawing, wh
pencilling draftsman, through the tr
engineer.

In addition to the above, there
the title, the shop order number, the
drawing in the case where the drawir
provided the machine is manufactured

160. The words indicating the nar
letters than the rest, since it should l
title.

The words indicating the type of
address, and the scale, should be cor
but they must be smaller than the l
the name of the machine.

The words used to indicate the r
objects as drawn, should be expressec
as, so many inches to the foot or so
if not, a proportionate size should be
etc.

The date and the **names** or initia
drawing should be in **very small lett**

All lines in the title must be ar
to a center line.

The title or border lines, or both,
with an ordinary printing-press, befoi

The title is occasionally **put in v**
the tracing in red ink; after which
ink. The black drawing ink shoulc
still wet.

161. In the lower left hand corn
and number for an **index mark** whic

MANUAL OF DRAWING.

and serial number of the drawing respectively; as, for example, **A–2** will indicate that the sheet is 36×48, and it is the second consecutive drawing that has been made.—**Plates I–V.**

162. The letters which should be used with the corresponding **sizes of sheets,** are as follows: **A,** 36×48; **B,** 24×36; **C,** 18×24; **D,** 12×18; **E,** 9×12; and **F,** 6×9.

163. **Bill of Material.**—Every detail drawing should have a bill of material which should be placed directly above the title and it is a good plan to block out a space for it immediately after the title has been allowed for.—**Plate II.**

Every piece in a machine, or its parts, or in any structure, should be **accounted for** on the bill of material, so that any clerk in the office can order from it independently of anything else.

In the **table of the bill of material,** the first column should contain an identifying mark which is exactly the same as the one placed on the view of the piece as shown on the drawing and may be a letter or a number. The second column should contain the name of the piece. The third column should contain the number wanted of the same piece for one machine, or one composite structure of any sort. The fourth column should contain the name of the material of which the piece is made. And the last column should contain any further necessary description of the piece, which may be; for example, the pattern number, the dimensions of the rough stock from which it is made, the method of casting, etc., and in fact the last column should provide for any description of the piece not found in the first four columns and which would be essential for completeness in the order for the stock.

The name of the piece in the bill of material should be a **common shop term** used by the mechanics, and if there is none, a simple and suggestive one must be used.

Every piece that is cast from a pattern should have, and be given, a **pattern number** in the column under the head of "Remarks."

164. **Views.**—The number of views is determined by that judgment which serves common sense.

All the views **necessary** and no more should be drawn.

The selection of a view or views which sh comprehensive manner should always be mad

165. In simple and symmetrical pieces, as nut, plain gears, etc., **one view** is sufficient.—D

166. If an object is symmetrical in every re show some irregularity, it is sometimes only ne **view.**—Plate V.

167. In a very crooked piece, **several views Fig. 11.**

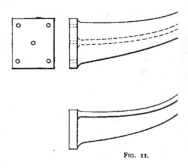

FIG. 11.

168. A **sectional** view is often clearer tha mechanic who works from the drawing but it s clearness obtains.

169. If an object is symmetrical, it is often a **combined outside and sectional** view; usual the center line.—**Fig. 12.**

170. All **bent** and **unusual shaped plates developed,** if it is necessary to show proper sp outline of cams, etc.—**Fig. 12.**

171. It is sometimes essential that an object, which is drawn on reduced scale, should have some important section of it shown full size.—Plate XII.

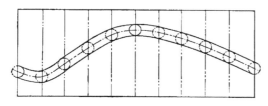

FIG. 12.

When there are two pieces that are exactly alike in every respect except that they are of opposite hands, one should be drawn in all of its views and a note should be added to indicate that they are *rights* and *lefts.*

172. In any view where there are a number of concentric circles, show only a few, as they are practically worthless for conveying any idea and only useful to fill space.

173. A flat surface on the body of a screw, stud, or shaft, is occasionally indicated, when there is only one view, by a rectangle with its intersecting diagonals.—Fig. 13.

174. Rectangles, with their intersecting diagonals, are often used for indicating the bearings on a round shaft; as, a line shaft, crank shaft, etc.—Fig. 14.

FIG. 14.

FIG. 13.

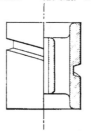

FIG. 15.

175. A knurled or milled piece should be shown conventionally in a view.—Fig. 15.

176. In all elevation views of squa one side.—Fig. 16.

FIG. 16.

177. In all elevation views of he three sides.—Fig. 17.

178. If, in a side sectional view description, the cutting plane does i them on a circle in another view, one shown actually cut by the plane and center of the circle.—Pipe Plug, Plate

179. Washers and collars should the piece to which they belong.—Driv

180. Keys should be shown in th Feed Screw, Plate II.

181. The side views of a Sellers V bastard or acme standard thread, kn

FIG. 18.

wood screw thread, are conventionally tive order.—Fig. 18.

182. If a **threaded piece** is **sectioned** nothing but the V's should represent the thread.—**Fig. 19.**

183. The **end views of a screw and tapped hole** are conventionally shown in the figure in their respective order.—**Fig. 20.**

FIG. 20.   FIG. 21.

184. A **filister head screw** should show, in all views of the end, its slot making an angle of 45° with the horizontal and the slot should show, in the side views, a true projection of the end views.—**Fig. 21.**

185. **Spur, bevel, worm, and spiral gears** are shown in detail.—Spur and Bevel Gears, Plate XVI; Worm Gear, Fig. 22; Spiral Gears, Fig. 23.

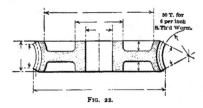

FIG. 22.

186. **Cast gears** made from a pattern should have, in addition to an ordinary view, one tooth completely detailed and the number of teeth specified.

187. The **arrangement of views** on the drawing demands, for the sake of clearness, system, and convenience: first, that they should not be crowded; second, that all views of the same piece should not be separated any further than necessary to show outside dimensions and notes clearly; third, that all views of different pieces be sufficiently separated to pre-

vent any confusion as to their proper relations to ·
fourth, as far as possible, all pieces of a machine m

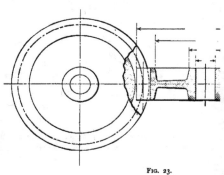

FIG. 23.

drawing in their natural relative positions in the ·
other.

188. **Sectional views** can be **placed**, if conver
any place on the sheet provided they are properly n

189. To cover the drawing sheet properly with ·
convenient to approximately **pre-arrange** the locat
enclosing them in a rectangle which is sufficiently
outside dimensions and notes to the view. At other
sary to do much planning, but **commence at** the up
and work across and down the sheet. This order ·
sheet should always be followed.

190. The relation of views to each other must al·
nically called **third angle projection** which, inter
when there is a definite and natural top to an object,
be the top view which is often called the plan; that
vation should be shown below the top view; and th

projection must be shown as the near side of the adjacent view from which it is projected.

191. **In a sectional view,** the portion of the object nearest the view is removed, and all of **what is left is shown in projection.**

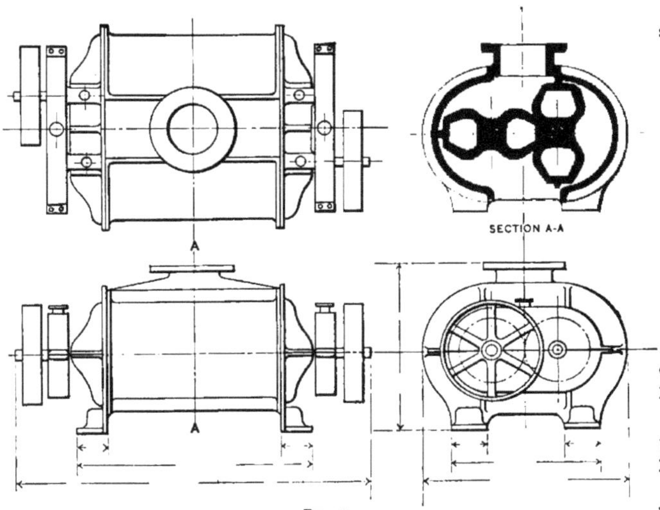

Always show **filletted or rounde** faces do not forbid.

194. All lines bounding a **hexa** to a circle whose diameter is equal of the hexagon and by means of th

195. When the three faces of shown, make the **distance between**

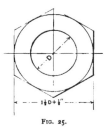

FIG. 25.

SECTION A-A

FIG. 24.

diameter of the body of the screw **lines** a trifle less than the diameter

196. Curves of intersection sho possible.—**Curve of intersection of IV.**

197. An **ellipse** in projection w by convention, by a circle.—**Fig. 26**

198. An **ellipse** whose **major** is drawn with circular arcs with radii of major axis, and the end arcs t through the ends of the major axis.

When the minor and major ax **true ellipse** proceed as follows. dr from a center at the intersection of

192. **Lines of Object.**—A **visible edge** of a solid should be represented by a full line, clean cut, comparatively wide to make it stand out in a bold effect, and of uniform width throughout its length.—**Plate I.**

193. Where a **corner** is not sharp, but rounded, it is admissible by convention to represent it in all views, if any clearness is gained thereby. —**Curve of intersection of pulley-arm with bead on rim, Plate IV.**

center of circles; draw horizontal and vertical lines through points of intersection of radial line and minor circle and of radial line and major

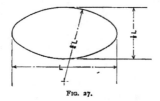

FIG. 27.

circle respectively. The points of intersection of the horizontal and vertical lines will be points on the ellipse.—Fig. 28.

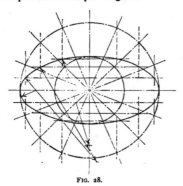

FIG. 28.

To draw **any ellipse with circular arcs** passing through predetermined points, proceed as follows: considering one quadrant, draw from a center on the major axis a circular arc which will pass through the end of the major axis and the first point; from a center on the line passing through the first point and the center of the first arc, draw a second arc

through the first point and second point; continu the points have circular arcs connecting them. arcs to a quadrant is usually sufficient for a very the ellipse. Care should be exercised in making ; of tangency of the arcs.—**Fig. 28.**

199. Make the **chamfers on bolts,** studs, shaf **21.**

200. **Oil holes and channels** are usually not s be made in an unusual manner.

201. **The spacing of V-thread lines,** and their lutely exact but a fine thread naturally has its l has less pitch than a coarse thread.

202. The **spacing of all thread lines, excep** approximately close and always done by markin from the scale.

203. Remember that **nuts,** when **sectioned,** pitch of the threads in reverse order to those on tl **Fig. 19.**

204. When a **screw** is of **considerable lengtl** should be drawn at the ends only, leaving the in —**Feed Screw, Plate II.**

205. An **invisible edge of a solid** should be r line, which is clean cut and composed of compa equal lengths. The dashes should be drawn as sible, and not quite so wide as the visible lines of

An **invisible** line, composed of **more than o** begin and end on the terminating lines which includ **Plate II.**

If an **invisible line** is shown by **one dash** only end at the terminating lines, but a short space tween the dash and terminating lines.—**Fig. 29.**

206. An **adjacent part line** is a line **in anoth** cent to the one regularly shown on the drawing purpose of indicating, for some particular reasc

the piece shown on the drawing. Therefore it should be a main outline, of sufficient length to indicate what is attached. It should be a fine, broken, clean-cut line composed of alternating dashes of one eighth of

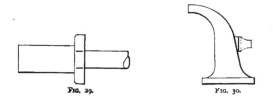

FIG. 29.   FIG. 30.

an inch and three eighths of an inch lengths respectively, drawn as close together as possible.—**Fig. 30.**

An **adjacent part line** is also used for **showing** some important **part** of an object that is **cut away** on the section view.—**Fig. 31.**

207. A **broken line** represents the visible edges of the part of a solid of a structure that is **broken off.** The width of this line should correspond

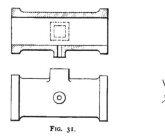

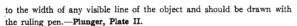

FIG. 31.   FIG. 32.

to the width of any visible line of the object and should be drawn with the ruling pen.—**Plunger, Plate II.**

208. An **alternate position line** is important position of a moving part the drawing. An alternate position ing the bare outlines of the object or lines are drawn, they should be fine of alternating dashes of one eighth of lengths respectively, drawn as close to

209. **Sectioning.**—Where there is cutting plane shown on the view of the sented by a fine line whose character is also to be notated at each end an be suitably placed and referred to, section.—**Fig. 24.**

210. **Section lines,** which are some parallel lines at equal distances fro represent a surface which has been c

Uusally **hatch lines** make an an it is not criminal, but even sometime if the conditions warrant it.

211. The **spacing of sectioned line** of the sectioned surface on the drawi is cut. But sectioned lines should no inch apart and softer material shoul ther apart than harder material.

Adjacent pieces shown **in section,** verse order.—**Plunger, Plate II.**

212. All **hatching** should be do which are to be inked.

When the surface is large, the hat **along the edges only.**

There are **conventional lines to re** are cut, but they are often unintel relegation to oblivion is forcibly rec represent babbitt is perhaps universal

213. A soft drawing or crayon pencil is often rubbed on the surface of the drawing to represent the section, but a blue crayon pencil should

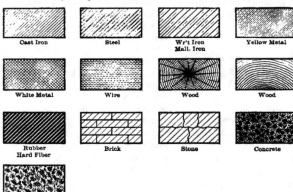

Cast Iron   Steel   Wr't Iron Mall. Iron   Yellow Metal

White Metal   Wire   Wood   Wood

Rubber Hard Fiber   Brick   Stone   Concrete

Earth

FIG. 33.

never be used for the purpose when a print is to be taken from it.

214. A rib, an arm of a pulley, or any comparatively thin piece,

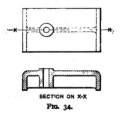

SECTION ON X-X
FIG. 34.

should not be sectioned by a cutting plane parallel to its largest bounding surfaces.—Fig. 34.

215. A solid round piece should not be sho cutting plane is taken through its axis.

216. A turned section placed on a view is and often clearer than when placed off at one Plate IV.

217. A keyway in a hole should not, as a should be shown by an invisible line.—Plate IV

218. Pieces that are broken at the ends a in the figure.—Fig. 35.

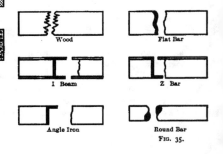

Wood   Flat Bar

I Beam   Z Bar

Angle Iron   Round Bar
FIG. 35.

219. Center Lines.—A center line is used to any configuration which has a natural axis or a lines to represent it or them. A circle, for ins center lines at right angles to each other.

Pitch lines of gears should be shown as cent

220. When circles are spaced along a circum of a disc, as bolt circles on a flange, a circular an ing through the bolt hole centers, are used as the

221. A center line should be a fine, broken, c of alternating dashes of one eighth of an inch a inch lengths respectively, evenly spaced and not short center line may consist of one dash.—Plate

222. Penciled center lines that are to be inked or traced over should be full lines.

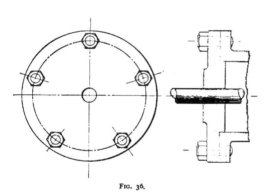

Fig. 36.

223. Visible and invisible lines of an object and surface shade lines have **precedence over center lines** in case there is coincidence.

224. Dimensions.—There is **nothing** on the whole drawing **so important as the dimensions.**

225. A **dimension line** is used for **indicating** a dimension between certain limits, and it is broken at some convenient place between its limits (preferably midway) to allow for the dimension.

226. The **character** of a dimension line should be a fine, broken, clean cut line, composed of equal dashes, equally spaced from each other.—**Plate I.**

227. All dimension lines, excepting those representing radii of circular arcs, are **terminated by arrow heads** at both ends. A **dimension line** which indicates a **radius** is terminated by an arrow head at the arc only, and if there are no intersecting center lines at the center from which the arc is struck, a small cross should indicate it. *Rad.* should be placed after the dimension.—**F:g. 37.**

228. The **arrow head** should make sion line, and should be made in sh(which is used drawing.

229. All **di** **reference to ma** 230. The p **determined by j** that view which manner as to what the dimen line stands for.

231. It is we piece over its vie

Fig. 37.

232. All the dimensions to any **pa** of a circle, rib, rectangular surface, e possible and *not* on several views.

233. A **dimension** is more clear a **itself**, provided it does not cause a con ness, and, provided there is sufficient figures between.

The **dimension,** however, may be witness line drawn from it to the usua sion line.

234. Dimension lines should not to impair clearness.

235. When **dimensions** are grou; dimension is inside and the others ar on the outside.—**Fig. 22.**

If one or the two **ends** of a dimen **lines** which are close together and th of the parallel lines limits the dimen limited by the outside one of the line should be used with arrow heads just t

236. A **short dimension** can be shown by placing the dimension lines with their arrow heads on the outside of the limiting lines and the dimension inside.—**Fig. 39.**

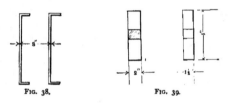

FIG. 38. FIG. 39.

237. If **one end** of a **dimension line cannot be shown** the full dimension is given at a break in the line.—**Fig. 40.**

A line of **sub-dimensions** should usually commence at the finished surface, and, if there is an unfinished surface on the opposite side, the

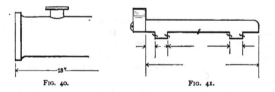

FIG. 40. FIG. 41.

last sub-dimension should be omitted, and, also, an **over all dimension** should be given from surface to surface.—**Fig. 41.**

238. The common two-foot rule has undoubtedly been generally adopted for a standard when dimensions are given in feet and inches. Therefore, all **dimensions** must be given in inches including twenty-four inches and below. Dimensions above twenty-four inches must be given **in feet and inches.**

When dimensions are given in feet and inches, a dash one eighth of an inch long should separate them.

Foot and inch marks not less than one sixteenth be used for accents.

If there is a **fractional number less than an in** is expressed in feet and inches, the **figure zero s** it.—**Feed Screw, Plate I.**

A fraction must **never have a diagonal divisi** must be a little **smaller than the integer** which pre

239. When **dimensions** are given in **decimals,** be placed between the whole number and fraction

If the **decimal dimension** is less than an inch recede it and the decimal point and the inch r as indicated above.

240. If the **outside diameter of a gear** is a frac in a **common fraction.**

241. All **dimensions** should **read from the t side** of drawing. If **dimensions** are **at an angle** sheet, they should **read from bottom side of the d**

242. If a **dimension** is **not to** the same **scale** indicate it by a dash above the dimension.

243. **Dimensions** on a drawing should **indica** independent of the scale of the drawing.

244. All the **dimensions** given on a piece in **sions** of the piece when it is complete and n made for the shrinkage of casting, etc.

245. Care and judgment must be exercised as diameters are the dimensions needed for the mech **should have** its **radius** given; **a bored hole, its di piece, its diameter** given, etc.

If in a **side view** a round or square shaft or s there is **no circle or square in the end view** to rep might be confusion as to which was intended, should **always be placed** after the dimension.—**Fee**

A **dimension** should **never be placed on a cente**

It is an inviolable rule to **never cross a dimen**

Small rounds on corners, or fillets, are not usually dimensioned.

Circles which do **not** indicate clearly their relation to the piece shown in the view should never be **dimensioned.**

246. If the **end of a screw,** stud, etc., is rounded, the **dimension** should be given to **the corner** and not to the extreme end.—**Stud, Plate II.**

247. If **bolts and screws** are drawn and everything is **standard** except their lengths, only the diameters of their bodies, the length under the heads, and their lengths of thread are shown in the dimensions.

248. **Bolt heads and nuts** (when not standard), and square and hexagonal figures should be **dimensioned across flats** and not across diagonals.—**Fig. 25.**

249. **Rolled structural steel** should be **dimensioned in the commercial sizes** and not in every detail.—**Fig. 42.**

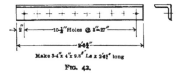

Make 3-4″ x 4″ x 9.8″ Ls x 2′6½″ long

FIG. 42.

250. Where a number of **rivet or small bolt holes** are in a right line, the usual dimensions should be given in a note on the dimension line between the first hole and the last hole. The note should give their number, size, distance apart and the total distance from the first to the last hole. The dimension from one of the end holes to the end of the piece should be regularly given.—**Fig. 42.**

251. The **sizes of pipes** must be stated in their nominal inside diameters.

252. **Pipe tapped holes** must be indicated by the size of pipe tap required and all other data regarding the hole is useless and should not be stated.

253. In giving **board dimensions,** remember that the thicknesses of stock boards are more likely to be in fractional than in even inches; as, a

seven eighths of an inch instead of
fourths inch instead of a two in
board dimensions.)

254. **Brick dimensions** should
for height.

255. **Windows** in brick walls sl
and **doors** should be **dimensioned**

256. When **angles** are given in
and minute accents should be used
replace the usual right line. Othe
the same rules as are observed for

257. **Dimensions of angles** on
are often more convenient when gi
ence line or point than when given

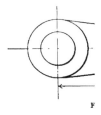

F

As a rule, the **dimension figure**
as the lines of the object.

258. **Extension lines are limiti**
is placed off the view and must
them equal to the same distance as
be fine, broken, parallel lines comp
distances, and no extension line sh

Short extension lines of one das

259. **Notes.**—Notes should be
be read easily and quickly.

If convenience admits of it, and there are several **words** in the note, they are more quickly read when placed **in a horizontal line.**

260. **All letters** used in notes on drawings must be of the **most simple type,** plain, of uniform height, and composed of lines of uniform width to match the projection lines of the object.

The **initial letter** of every word, excepting prepositions, should be taller than the others.

All the **letters in a word should be close together** and not less than one sixteenth of an inch nor more than one eighth of an inch in height.

261. **Draw parallel guide lines** in pencil before printing words. All words should be sufficiently separated for clearness.

262. **All notes** should be **written on pencil drawings** which are to be traced, and **printed on inked or finished** pencil drawings.

263. **Notes** must not be too **abbreviated.** '

All notes must be in **short sentences, explicit and concise.**

264. **Castings** and **rough forgings** are occasionally not finished all over; as a consequence, the custom in practice is to note the surfaces on each piece which are finished.

When a surface is machined or finished, an ƒ mark should be placed on that line of the view which represents it and not on the view where the blank surface only is shown.—**Plate V.**

If a casting is **finished all over** the several ƒ marks should be left off and the words *fin. all over* should be printed under or near the views.

265. When there are **two lines close together** representing the edges of surfaces, and one or both of them represent finished surfaces, the ƒ mark should be placed outside with a witness line drawn to the finished surface.—**Fig. 44.**

266. It is sometimes clearer, or necessary, to **state the kind of finish** or machining a piece should have, as *file finish, grind, plane, bore,* etc.

Pieces which are **made from rolled** or wrought steel bars, as spindles, shafts, studs, etc., are usually finished all over; and it is the custom to omit the finish notes.

267. If a **steel casting** is wanted do not confound it with cast steel but call it a *steel casting* everywhere on the drawing.

268. A plane surface that is **partially finished** with a dimension. If it is **counterbored** to furnis etc., only the note *spot face* is necessary.—**Fig. 45.**

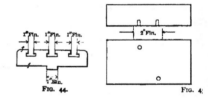

FIG. 44.   FIG. 4

269. If in machining a **fixture or jig** is used fo part in a machine, it should be indicated on or near

270. **Threads** should be **noted in number per** U. S. standard; in which case, they should be indic

271. **Tapers** on shafts should be given in inches

272. **Springs** are described in detail by giving a coil, the number of free coils to the inch, and th mensions of the body of the spring.

273. The **mesh of wire** in screens is indicated gage used, and the number per inch or the amou the wires.

274. **Forced fits** should be indicated by the t per unit area.

275. The **inclination of a pipe or roof** should pitch or slope and given in the ratio of the unit ri zontal distance; as, 1 in 6 pitch or one sixth pitch,

276. **Every piece** shown on a drawing must **hav** which should be a capital letter somewhat larger in the other notes on the drawing and must be th in the bill of material for the same piece. The **must not be used,** and, when the letters in the al

small capitals should be used along with and at the side of the large capitals.—Plate I.

277. **Shading.**—Shade lines for outline shading are sometimes used on edges and surfaces of views which conventionally cast shadows. They should not be used too freely, for they are a convention which adds to the work of the draftsman, but, occasionally, they make clearer the configuration ot an intricate piece which is shown on a drawing, provided that the mechanic understands the convention as well as the drafts. man who uses it. They are occasionally used for the purpose of display, which, at its best in detail drawing, is very much out of place. If, however, they are used on the views of one piece, they should be used on all the views of all the pieces which are shown on that one drawing in order to conform to that fundamental rule of uniformity which should obtain on all drawings.

278. **Outline shading** should be done by drawing a second adjoining projection line on and not outside that part of the view which represents the surface of the material. By drawing a shade line then, on the surface of the piece, it **cannot be considered** as **a line in the shadow** which is cast by an edge of an object as it is often interpreted.

By considering a **shade line** as a line in the shade of the piece and **not as a shadow** cast by it, the actual dimensions from edge to edge are kept to scale; and it follows, that the **lower** and **right hand outside edges** of a piece and the **upper** and **left hand edges** of holes would have shade lines.

279. When a **circle is shaded,** the set of the compass which strikes the circle should not be changed, but another center should be taken up and along a line which makes 45° with the horizontal. The eccentric arc which is then struck with the new center to represent the shade line should just join the original circle at the end of the 45° line and should continue until it just merges with the original circle.

280. In the side view of a **round piece** there is theoretically no outline shade line on the round edge but, by convention, a shade line should be used.

281. An **unsectioned round pie** meridian plane has conventionally

282. In a **V-thread screw,** b thread line is shown as a shade li

283. In a sectional **V-thread** tional thread line is shown as a sh

284. **Pencil drawings** that are **be outline shaded.**

285. **Surfaces** of pieces, whicl **shade lined.** Surface shading sho on commercial mechanical drawi sary to make clear or stand out s wholly confused in the view of an

As a rule, **no surface shading** The exception to this rule might b

286. **Penciled Drawings.**—W with all its views, should be compl penciling should be done in the fc

1. Draw **border lines.**

2. Draw **match lines.**

3. Block out **space for titles.**

4. Block out **space for bill of** n

5. Draw **main center lines** of o

6. Draw **main lines of object.**

7. Draw **small** and **inside lines** which shows the most essential fea

8. Put in **dimensions** and nece

As large **arcs** as possible should

It is an art to **make** a finishe **properly.** With some draftsmen it it must be acquired.

287. A **drawing,** which is to **completely penciled.** It is an und

sary tax on the mind, when a draftsman leaves a large share of the details in his mind for inking, instead of penciling them on the drawing paper.

288. **Inked Drawings.**—All drawings must be **inked** in with **black waterproof drawing ink.**

289. The different parts of a **drawing** should be **inked** in the following order: first, all the lines of the object, which are represented by the **small arcs**, the **large arcs**, and **right lines** taken in their respective order—and if there is any outline shading it should be done at the same time; second, all **auxiliary lines excepting hatch lines**; third, **dimension lines, arrow heads** and **notes;** fourth, **hatch lines;** fifth, **surface shade lines,** if there are any; sixth, **bill of material;** seventh, **title;** eighth, **border lines.**

290. It is sometimes well to ink only a **section** of a drawing at a time, particularly if the drawing is not to be finished the same day that it is begun.

It is rather imperative that **tracing cloth** should be **inked** in sections, as change of atmospheric conditions has a marked effect on the tracing cloth.

291. Always **commence** the **inking at the top and left-hand side** and **work across and down** the sheet.

292. **Any line** which is over one thirty-second of an inch **wide,** and drawn with a medium size ruling pen, should be made with more than one stroke of the pen.

293. When **inking** in the sections of **very narrow walls** in building plan or like sections, draw heavy parallel lines to represent their outlines and fill in with the writing pen or, preferably, a large ruling pen.

294. **Hatch lines** which are **three sixteenths of an inch long** or less, should be put in with the writing pen.

295. That which makes an inked drawing look well is **uniformity in every detail,** to wit: All figures, letters, lines, etc., that are used for like purposes should be of the same height, same width of line and same style.

296. **Checking Drawings.**—Everything on a drawing must be thoroughly checked by at least two persons before it is allowed to pass into the shops.

A draftsman should **check** his drawing **by** methodical **steps** and

should not try to check it by taking a bird's-e
done.

The following steps are to be observed in che

1. **Identify every piece** in its relation to the n
of proper form and that none are missing.

2. Check every view for **correct and complete**

3. Note if there are any required **dimensions**
ing.

4. **Scale every dimension,** and note if they a
on the drawing.

5. **Check** the **main dimensions** of the pieces
and note if they agree.

6. **Check** the **arrow heads** to see if there are a

7. **Check** the **accents** to figures as inch marks,
marks.

8. **Check** the **center lines.**

9. **Check** the supplementary **notes** and **marks**
rect and none missing.

297. **Assembly Drawings.**—The function of
to show the arrangement of its principal parts in
and comprehensive manner.

298. An arrangement of parts is sometimes
part has its outline **fringed** with a **dotted section li**
ever is not common at the present time.—**Stub
fourth grade.**

299. On a small scale, it sometimes happens t
projection would be **close together** and even mer
is the case, the lines should be spaced a little fa
then one left out.—**Fig. 24.**

300. In assembly drawings show **elevations in**

301. **Minor details,** as nuts, keys, set-screws,
the assembly drawing when they are evident with

302. **Over all and important main dimension**
needed in an assembly drawing.—**Fig. 24.**

303. **Structural steel parts,** when they are shown **on** such a **reduced scale** as to make a section one eighth of an inch or less in width, should be blacked in.—**Fig. 35.**

304. When there are four or more **bolts** in a piece, two should generally be shown, in a side view, at their true scale distance from the center, as measured each side of the center line.—**Fig. 36.**

305. Show only **pitch circles** in **end views of gears.**—**Fig. 46.**

306. **Diagram Drawings.**—If merely an outline of a machine is

307. **Patent Office Drawings.**- made to certain specifications whic the official "Rules of Practice" in t

When the **invention** consists of the drawing must exhibit, in one o connected from the old structure, only of the old structure as will s invention therewith.

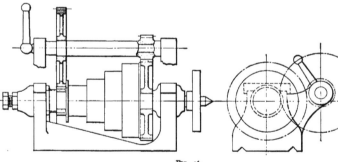

FIG. 46.

desired, a diagram drawing is made, which is composed of the main lines of the machine with the usual center lines.—**Fig. 47.** If the **relation of** some of the **moving parts** of a machine is desired, they may be represented by connected heavy lines passing through the center line of the members, as a Corliss valve gear, etc.—**Fig. 48.**

In some **intricate types** of machinery, as shoe machinery, textile machinery, etc., and especially in textile machinery, it is often quite impossible to show, for an assembly drawing, anything more than a diagram drawing.

Drawings must be **made upon** responding to three sheet **Bristol b** be calendered and smooth. India i fectly black and solid lines. (Tw drawing ink is used a great deal b

The **size of a sheet** on which 10×15 inches. One inch from its drawn, leaving the "sight" precisel all work and signatures must be in

the sheet is regarded as its top, and, measuring downwardly from the marginal line, a space of not less than $1\frac{1}{4}$ inches is to be left blank for the heading of title, name, number, and date.

to all lines, however fine, to shading, and to li
faces in sectional views. All lines must be clea
they must not be too fine or crowded. Surfa
should be open. Sectional shading should be r
lines, which may be about one twentieth of an i
should not be used for sectional or surface shadir

Drawings should be made with the fewest
with clearness. (This means no center lines,

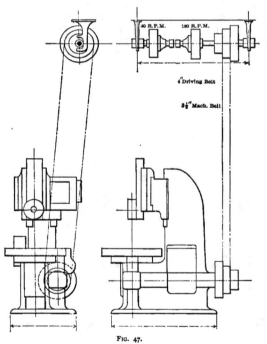

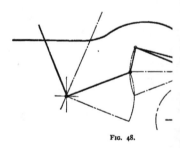

FIG. 47.

FIG. 48.

All drawings must be made with the pen only. Every line and letter (signatures included) must be absolutely black. This direction applies

Author.) By the observance of this rule the ei
after reduction will be much increased.

Shading (except on sectional views) should t
and concave surfaces, where it should be used
even there dispensed with if the drawing is otherv
plane upon which a sectional view is taken sho
general view by a broken or dotted line. Heavy
of objects should be used, except when they te:
and obscure letters of reference. The light is al
from the upper left-hand corner at an angle of 4
or surface graining should not be attempted.

The scale to which a drawing is made ough
show the mechanism without crowding, and two

be used if one does not give sufficient room to accomplish this end; but the number of sheets must never be more than is absolutely necessary.

The different views should be consecutively numbered. Letters and figures of reference must be carefully formed. They should, if possible, measure at least one-eighth of an inch in height, so that they may bear reduction to one twenty-fourth of an inch, and they may be much larger when there is sufficient room. They must be so placed in the close and complex parts of drawings as not to interfere with a thorough comprehension of the same, and therefore should rarely cross or mingle with the lines. When necessarily grouped around a certain part, they should be placed at a little distance where there is available space, and connected by short broken lines with the parts to which they refer. They must never appear upon shaded surfaces, and when it is difficult to avoid this a blank space must be left in the shading where the letter occurs, so that it shall appear perfectly distinct and separate from the work. If the same part of an invention appears in more than one view of the drawing it must always be represented by the same character; and the same character must never be used to designate different parts.

The signature of the inventor should be placed at the lower right-hand corner of each sheet, and the signatures of the witnesses at the lower left-hand corner, all within the marginal line, but in no instance should they trespass upon the drawings. The title should be written with pencil on the back of the sheet. The permanent names and title will be supplied subsequently by the office in uniform style.

When views are longer than the width of the sheet, the sheet should be turned on its side, and the heading will be placed at the right and the signatures at the left, occupying the same space and position as in upright views, and being horizontal when the sheet is held in an upright position; and all views on the same sheet must stand in the same direction.

As a rule, one view only of each invention can be shown in the Gazette illustrations. The selection of that portion of a drawing best calculated to explain the nature of the specific improvement would be facilitated and the final result improved by the judicious execution of a figure with

express reference to the Gazette, bu
as one of the figures referred to in
the figure may be a plan, elevation,
to the judgment of the draftsman.
good substitute for the true perspec
establishing three axes which theore
other; of drawing lines parallel to t
object; and of making them either
an axometric drawing is used for a w
ing should be to a full or some conv
is desired, certain lines may be slight

The arrangement of the axes is
obtained in drawing the lines of the
convenient arrangements may be
as, Isometric, Cavalier, and Cabine
drawn with axes making 120° with
other two making 30° with the hori
with one of the axes making any
two being vertical and horizontal.
one of the axes 45° with the horiz
and vertical. Isometric projection
cabinet projection usually gives the

In making an axometric projec
drawing may be facilitated by draw
around plan and elevation of object
the vertical and 30° axes correspon
the enveloping prism and to the tw
base; that, if it is cavalier, the ver
correspond respectively to the vertic
to the two right angle edges in its ho
vertical and horizontal and 45° axes
edge of the enveloping prism and to
zontal base; and also that lines in th
axis are shortened to one half their a

of orthographic projection, lines which are parallel in the object will be parallel in the drawing; and that all irregular lines may be drawn if their connecting points are first formed. Isometric, cavalier and cabinet cubes are shown in **Fig. 49.**—Author.)

It must **not cover a space exceeding 16 square inches.** All its **parts** should be **open and distinct,** with very little or no **shading,** and it must illustrate the invention claimed only, to the exclusion of all other details. When well executed, it will be used without curtailment or change, but any excessive fineness, or crowding, or unnecessary elaborateness of detail will necessitate its exclusion from the Gazette.

An **agent's or attorney's stamp, or advertisement, or written address**

will **not be permitted** upon the face of a drawing

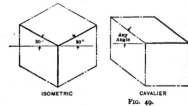

ISOMETRIC   CAVALIER
FIG. 49.

marginal line.—**Plate XXI.**

INDEX.

PLATE V.—GEAR.

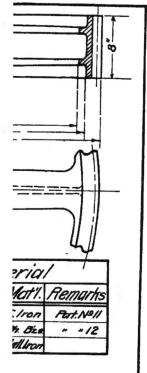

8″

rial		
	Mat'l.	*Remarks*
	:Iron	*Part.Nº11*
	½ Size	*" " 12*
	d.Iron	

lf-Oiling Bushing

hley College

New York

I ft.

Name
11

PLATE V.

GEAR.

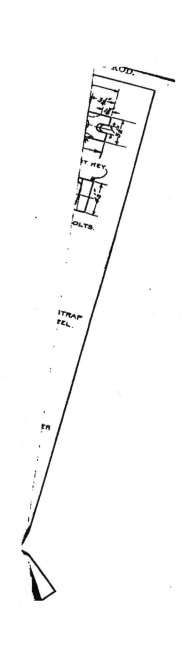

ID.

v#355

S
VD
ETC.
Q.
A.

ION.

EMPLET
ENGINE
COLLEGE
W YORK
T.
        NAME

SHAPER.

68° 51'

65°

26° 34'

Gear.

ne Shaft

Bushings

PA'

www.ingramcontent.com/pod-product-compliance
Lightning Source LLC
LaVergne TN
LVHW012202040326
832903LV00003B/60